Monsters Galore:
A Fun Counting Adventure

A MISTER JON BOOK

BAD COOKING

ISBN: 9798394002533
IMPRINT: BAD COOKING

FOR ZACH

Dear Parents,

Welcome to "Monsters Galore", a colorful and fun-filled counting book for young children. As a parent, you know the importance of providing your child with the best possible tools to learn and grow. "Monsters Galore" is designed to help your child develop early math skills while having fun with a playful cast of friendly monsters.

In this book, your child will learn to count from 1 to 25 through engaging illustrations of monsters in common public places such as the park, the beach, and the grocery store. Each page offers a new adventure, and your child will have the opportunity to practice counting and identifying numbers in a fun and interactive way.

Not only does "Monsters Galore" help your child develop early math skills, but it also promotes visual learning, memory retention, and cognitive development. Your child will have a blast with the silly and friendly monsters while also learning valuable skills for the future.

We hope that you and your child enjoy reading "Monsters Galore" as much as we enjoyed creating it. Let the counting adventure begin!

Best regards,
Mister Jon

PLAYFUL AND FUN LEARNING CAN SPARK A LIFELONG LOVE OF EDUCATION AND BUILDS A STRONG FOUNDATION FOR YOUR YOUNG ONES SUCCESS.

LEVEL ONE
WARM UP

ONE

TWO

THREE

FOUR

FIVE

LEVEL TWO
QUICK ROUND

One silly monster

2

Two silly monsters,
counting as we go.

3

Three silly monsters

4

Four silly monsters,
on the counting flow.

Five silly monsters

6

Six silly monsters,
let's count them with glee

7

Seven silly monsters

8

Eight silly monsters,
counting monsters, whee!

Nine silly monsters

And then there were ten,
the numbers are climbing up again,

Eleven silly monsters,

Twelve and counting more,
monsters everywhere, let's explorer,

13

With thirteen monsters,

Then fourteen more,
we're getting closer to the score,

15

Fifteen monsters, we're almost through, counting with monsters, what fun to do!

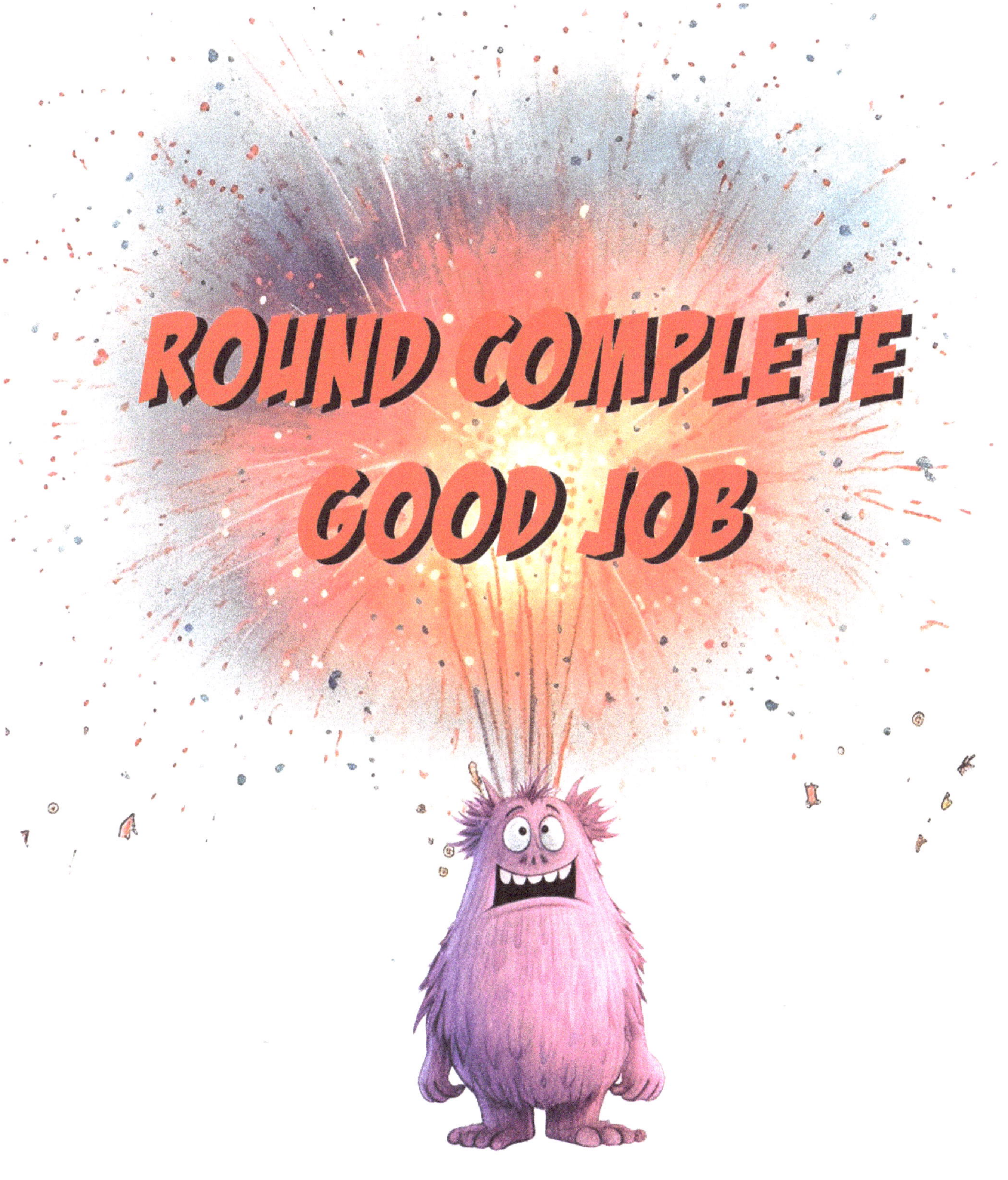
ROUND COMPLETE
GOOD JOB

1 One monster, big and green,
At school, it's the only one seen!

2 Two more monsters, red and blue,
In the airport, waiting to go

3 Three little monsters, happy and bright,
At the supermarket, quite a sight!

6 Four, five, six, all in a row,
In the park, where they love to go!

7 Seven monsters, ready to skate,
Quick RUN, before it's too late!

10 Eight, nine, ten, playing in the sand,
At the beach, it's just so grand!

11 Eleven mischievous monsters, having fun,
On the playground, under the sun!

13 Twelve, thirteen, hiding in the trees,
In the forest, where they feel so free!

14 Fourteen monsters, hopping with glee,
On the lawn, bursting with energy!

16 Sixteen monsters braved the cold snow,
As the penguins waddled in a perfect row.

17 Seventeen monsters, having a ball,
At the circus, where they're the stars of all!

19 Eighteen, nineteen, part of a show,
In the theater, where the lights glow!

20 Twenty monsters, with big wide eyes,
At the museum, where they're a surprise!

22 Twenty-one, twenty-two, reading a book,
In the library, where they love to look!

23 Twenty-three monsters, lined up in a row,
At the concert hall, where music's all aglow!

25 Twenty-four, twenty-five, ready to explore,
In the jungle, where wild creatures roar!

Great job counting, you did it with ease,
With monsters everywhere, it's just a breeze!
In familiar scenes, they're so much fun,
Counting monsters, now the adventure's done!

BUT not just yet!

BONUS ROUND

WHERE IS?

Can you find the 3 monsters in the crowd?

Can you find the four monsters?

KNOW YOUR MONSTER

MUFFIN C. CUPCAKESTEIN
HOBBIES: BAKING CUPCAKES, PLAYING
HOPSCOTCH, AND PRACTICING MAGIC TRICKS
FAVORITE FOOD: CUPCAKES AND MILKSHAKES
FAVORITE COLOR: BUBBLEGUM PINK

SNUFFLES P. SNICKERDOODLE
HOBBIES: PLAYING DRESS-UP, HAVING
TEA PARTIES, AND READING BOOKS
FAVORITE FOOD: COOKIES AND
HOT CHOCOLATE
FAVORITE COLOR: LAVENDER

ZIPPY Z. LIGHTNINGBOLT
HOBBIES: RUNNING RACES, JUMPING ROPE,
AND PLAYING TAG
FAVORITE FOOD: CARROTS AND APPLE JUICE
FAVORITE COLOR: THUNDERSTORM GRAY

SQUIRT K. JELLYBEANER
HOBBIES: SPLASHING IN PUDDLES,
PLAYING WITH WATER GUNS, AND
FISHING FOR RUBBER DUCKS
FAVORITE FOOD: JELLY BEANS AND
LEMONADE
FAVORITE COLOR: OCEAN BLUE

ZIPPY K. MCZIGZAG
HOBBIES: RACING, ROLLERBLADING,
AND JUMPING ON TRAMPOLINES
FAVORITE FOOD: POPCORN AND HOT DOGS
FAVORITE COLOR: BRIGHT YELLOW

GOOBER MCSILLYPANTS
HOBBIES: TELLING JOKES, PLAYING
PRANKS, AND MAKING PEOPLE LAUGH
FAVORITE FOOD: PEANUT BUTTER AND
JELLY SANDWICHES AND PIZZA
FAVORITE COLOR: SILLY GREEN

SPIKE Q. BALLER
HOBBIES: PLAYING SPORTS, ROCK
CLIMBING, AND EXPLORING
FAVORITE FOOD: PIZZA AND TACOS
FAVORITE COLOR: RED

TOOTS C. MCTOOTERSON
HOBBIES: PLAYING THE TRUMPET,
SINGING, AND DANCING
FAVORITE FOOD: CHOCOLATE CAKE WITH
SPRINKLES
FAVORITE COLOR: ROYAL BLUE

DOODLE J. MCSCRIBBLE
HOBBIES: DRAWING, DOODLING, AND
CREATING ART
FAVORITE FOOD: SKITTLES AND
GUMMY BEARS
FAVORITE COLOR: RAINBOW BRIGHT

BOUNCY MCJUMPERSON
HOBBIES: BOUNCING ON TRAMPOLINES,
PLAYING HOPSCOTCH, AND JUMPING ROPE
FAVORITE FOOD: JELLYBEANS AND LOLLIPOPS
FAVORITE COLOR: BRIGHT PURPLE

ZIPPY Q. LIGHTNING
HOBBIES: RACING, JUMPING, AND
FLYING KITES
FAVORITE FOOD: ENERGY BARS
AND SMOOTHIES
FAVORITE COLOR: YELLOW

ZIPPY K. MCFUZZ
HOBBIES: RACING AROUND, PLAYING
PRANKS, AND COLLECTING SHINY THINGS
FAVORITE FOOD: PIZZA WITH EXTRA
CHEESE AND PEPPERONI
FAVORITE COLOR: ELECTRIC BLUE

FLUFFY P. MCCLOUD
HOBBIES: SINGING, DANCING, AND
PLAYING MUSICAL INSTRUMENTS
FAVORITE FOOD: ANGEL FOOD CAKE
AND WHIPPED CREAM
FAVORITE COLOR: BLUE

DOOZIE MCFLAPPER
HOBBIES: FLYING, SOARING, AND
AERIAL ACROBATICS
FAVORITE FOOD: HONEY AND WAFFLES
FAVORITE COLOR: PURPLE

SQUISHY MCSQUASHY
HOBBIES: PLAYING IN MUD PUDDLES,
SPLASHING IN WATER, AND
GETTING MESSY
FAVORITE FOOD: POPSICLES AND
WATERMELON
FAVORITE COLOR: MUDDY BROWN

WIGGLES MCSQUIGGLE
HOBBIES: DRAWING, PAINTING, AND
SCULPTING WITH CLAY
FAVORITE FOOD: RAINBOW SPRINKLES
AND CHOCOLATE CHIP COOKIES
FAVORITE COLOR: SKY BLUE

GIGGLES MCLAUGHALOT
HOBBIES: TELLING JOKES, PLAYING
GAMES, AND MAKING FUNNY FACES
FAVORITE FOOD: POPCORN AND
COTTON CANDY
FAVORITE COLOR: SUNSHINE YELLOW

CRUNCHY CRACKER
HOBBIES: SOLVING PUZZLES, PLAYING
CHESS, AND STRATEGIZING
FAVORITE FOOD: PRETZELS AND
CHEESE CRACKERS
FAVORITE COLOR: BROWN

WHISKERS MCWHISKERTON
HOBBIES: CHASING MICE, SCRATCHING
TREES, AND TAKING LONG WALKS
FAVORITE FOOD: TUNA FISH AND CREAM
FAVORITE COLOR: MELLOW YELLOW

WALLY MCWALLACE
HOBBIES: BUILDING FORTS, SOLVING
PUZZLES, AND READING BOOKS
FAVORITE FOOD: PEANUT BUTTER
AND JELLY SANDWICHES
FAVORITE COLOR: DEEP GREEN

BOUNCY SPRING
HOBBIES: JUMPING ON TRAMPOLINES, BOUNCING BALLS, AND PLAYING HOPSCOTCH
FAVORITE FOOD: JELLO AND GUMMY BEARS
FAVORITE COLOR: GREEN

FIZZLE BUMBLEBOOM
HOBBIES: MIXING CRAZY POTIONS, EXPERIMENTING WITH EXPLOSIONS, AND PLAYING WITH FIZZY DRINKS
FAVORITE FOOD: BUBBLY SODA
FAVORITE COLOR: SPARKLING BLUE

BLINKY WOBBLEBOTTOM
HOBBIES: DANCING THE WOBBLE, PLAYING PRANKS, AND TELLING JOKES
FAVORITE FOOD: SPAGHETTI WITH EYEBALL SAUCE
FAVORITE COLOR: NEON GREEN

WOBBLE GIGGLESNORT
HOBBIES: BALANCING ON ONE FOOT, SINGING KARAOKE, AND TICKLING FUNNY BONES
FAVORITE FOOD: JELLY SANDWICHES AND GIGGLY GELATIN
FAVORITE COLOR: SILLY GREEN

WHIZZLE SNORTSNIFFLE
HOBBIES: SNIFFING FLOWERS, CHASING
BUTTERFLIES, AND DOING SOMERSAULTS
IN THE GRASS
FAVORITE FOOD: MARSHMALLOW AND
ANCHOVY SANDWICHES WITH EXTRA MUSTARD
FAVORITE COLOR: SNIFFLE PINKIPSUM

ZIGZAG BUMBLESNORT
HOBBIES: PAINTING SQUIGGLY LINES,
ROLLER SKATING, AND BOUNCING ON
POGO STICKS
FAVORITE FOOD: TWISTY PRETZELS
AND ZIGZAG-SHAPED COOKIES
FAVORITE COLOR: ZANY YELLOW

GIGGLES BUMBLEBOUNCE
HOBBIES: TELLING KNOCK-KNOCK JOKES,
TICKLING FUNNY BONES, AND BOUNCING
ON TRAMPOLINES
FAVORITE FOOD: GIGGLY GRAPES AND
JOLLY JELLY
COLOR: GIGGLETASTIC YELLOW

BOP JIGGLYTOES
HOBBIES: BOPPING TO THE BEAT, MAKING
SILLY NOISES, AND PLAYING AIR GUITAR
FAVORITE FOOD: BOUNCY JELLYBEANS
AND WIGGLY GELATIN
FAVORITE COLOR: BOP-A-DOODLE ORANGE

SWOOSH MCAIR
HOBBIES: FLYING, GLIDING, AND SKYDIVING
FAVORITE FOOD: MARSHMALLOW CLOUDS AND COTTON CANDY
FAVORITE COLOR: SKY BLUE

FIZZLE MCSPARKLES
HOBBIES: EXPERIMENTING, CREATING INVENTIONS, AND BUILDING CONTRAPTIONS
FAVORITE FOOD: PRETZELS AND ROOT BEER
FAVORITE COLOR: SILVER

SQUIGGLE MCLINE
HOBBIES: DRAWING, PAINTING, AND MAKING UP STORIES
FAVORITE FOOD: RAINBOW LOLLIPOPS AND FRUIT SALAD
FAVORITE COLOR: COLORFUL RAINBOW

GOOBER MCFUZZBALL
HOBBIES: PLAYING HIDE-AND-SEEK, EXPLORING, AND MAKING SLIME SCULPTURES
FAVORITE FOOD: JELLY BEANS AND PIZZA
FAVORITE COLOR: NEON GREEN

SQUISHY GUMMYBEARINGTON
HOBBIES: BOUNCING ON TRAMPOLINES,
PLAYING HIDE AND SEEK,
WATCHING CARTOONS
FAVORITE FOOD: GUMMY BEARS AND
ICE CREAM SUNDAES
FAVORITE COLOR: NEON GREEN

BOUNCY WOBBLEKINS
HOBBIES: JUMPING ON CLOUDS,
TRAMPOLINE BASKETBALL, AND
BLOWING BUBBLEGUM BUBBLES
FAVORITE FOOD: WATERMELON PIZZA
WITH CHEESE AND PICKLE TOPPINGS
FAVORITE COLOR: SKY BLUE

FIZZ WHIZZBANG
HOBBIES: MIXING CRAZY POTIONS,
RIDING UNICYCLES, AND TELLING
KNOCK-KNOCK JOKES
FAVORITE FOOD: MARSHMALLOW
SPAGHETTI WITH JELLY AND MUSTARD
FAVORITE COLOR: BUBBLY PINK

GIZMO WACKYTOOTH
HOBBIES: INVENTING CONTRAPTIONS
SOLVING PUZZLES, AND DOING MAGIC
TRICKS
FAVORITE FOOD: BANANA PEPPERONI
PIZZA WITH CHOCOLATE SAUCE
FAVORITE COLOR: ZANY ORANGE

WOBBLE NOODLEHEAD
HOBBIES: BALANCING PLATES ON HIS HEAD, DOING SOMERSAULTS, AND CHASING FIREFLIES
FAVORITE FOOD: SPAGHETTI WITH PEANUT BUTTER AND JELLY SAUCE
FAVORITE COLOR: SLINKY GREEN

FUZZY MCTICKLEBOTTOM
HOBBIES: TICKLE FIGHTS, BOUNCING ON TRAMPOLINES, AND JUGGLING BANANAS
FAVORITE FOOD: MARSHMALLOW PIZZA WITH PICKLES
FAVORITE COLOR: POLKA DOT PURPLE

BLINKY TICKLEFEET
HOBBIES: PLAYING HIDE-AND-SEEK, TICKLING TOES, AND SWINGING FROM TREE BRANCHES
FAVORITE FOOD: JELLYBEAN TACOS WITH RAINBOW SPRINKLES
FAVORITE COLOR: GIGGLY YELLOW

BUMBLE FUMBLEBUM
HOBBIES: BUMBLING THROUGH MAZES, PLAYING WITH BLOCKS, AND HIDE-AND-SEEK
FAVORITE FOOD: SILLY HONEYCOMB
FAVORITE COLOR: BUMBLEBEE STRIPES

SPRINKLE GIGGLEBERRY
HOBBIES: SPRINKLING MAGIC
GIGGLES, HAVING TEA PARTIES
WITH IMAGINARY FRIENDS,
AND PLAYING WITH CONFETTI
FAVORITE FOOD: GIGGLY CUPCAKES
FAVORITE COLOR: RAINBOW SPRINKLE

WACKY TUMBLETOES
HOBBIES: DOING SILLY DANCES,
TUMBLING DOWN HILLS, AND PLAYING TAG
FAVORITE FOOD: WOBBLY JELLY
FAVORITE COLOR: TOPSY-TURVY PINK

SNICKER WOBBLETOOTH
HOBBIES: TELLING SILLY STORIES,
COLLECTING FUNNY HATS, AND
PLAYING IN THE MUD
FAVORITE FOOD: SNICKERING
MARSHMALLOWS
FAVORITE COLOR: MUDDY BROWN

FUZZY WIGGLESWORTH
HOBBIES: GROWING WILD HAIRDOS,
PLAYING HIDE-AND-SEEK, AND DANCING
IN PUDDLES
FAVORITE FOOD: FUZZY PEACHES
AND SQUISHY MARSHMALLOWS
FAVORITE COLOR: FIZZLING PINK

SNORF SNICKERDOODLE
HOBBIES: COLLECTING SHINY ROCKS,
BUILDING SANDCASTLES, AND TAKING
BUBBLE BATHS
FAVORITE FOOD: MARSHMALLOW PIZZA
FAVORITE COLOR: BUBBLEGUM PINK

LISA GIGGLESNORT
HOBBIES: ROLLER-SKATING, PLAYING
PRANKS, AND TELLING FUNNY JOKES
FAVORITE FOOD: PICKLE AND JELLY
SANDWICH
FAVORITE COLOR: POLKA DOT RAINBOW

GOOBER BUMBLEBUM
HOBBIES: MAKING SILLY FACES,
PLAYING PRANKS, AND COLLECTING
SQUISHY TOYS
FAVORITE FOOD: PANCAKE TACOS WITH
MARSHMALLOW SYRUP
FAVORITE COLOR: SILLY STRAWBERRY

ZOODLE MCTICKLETOES
HOBBIES: HOPPING ON POGO STICKS,
PLAYING TAG, AND PAINTING RAINBOW
MURALS
FAVORITE FOOD: PEANUT BUTTER AND
JELLYFISH SANDWICH
FAVORITE COLOR: SUNSHINE YELLOW

WOBBLE SNICKERDOODLE
HOBBIES: BALANCING ON
TIGHTROPES, JUGGLING BANANAS,
AND READING UPSIDE DOWN
FAVORITE FOOD: WATERMELON
PIZZA WITH JELLYFISH TOPPINGS
FAVORITE COLOR: GIGGLE GREEN

SQUIGGLE WOBBLEBOTTOM
HOBBIES: TWISTING INTO PRETZEL
SHAPES, PLAYING HIDE-AND-SEEK,
AND SINGING SILLY SONGS
FAVORITE FOOD: SPAGHETTI ICE
CREAM WITH CHOCOLATE SAUCE
FAVORITE COLOR: RAINBOW SWIRL

BOGGLE WOBBLETOOTH
HOBBIES: BOUNCING ON
TRAMPOLINES,SPLASHING IN
PUDDLES, AND SOLVING
SILLY RIDDLES
FAVORITE FOOD: BUBBLEGUM SOUP
WITH JELLYBEAN TOPPINGS
FAVORITE COLOR: BOUNCY BLUE

GIZMO BUMBLEBEEBLE
HOBBIES: BUILDING LEGO TOWERS,
FLYING KITES, AND CREATING SILLY
DANCE MOVES
FAVORITE FOOD: CHEESEBURGER
CUPCAKES WITH FRIES
FAVORITE COLOR: ROBOT SILVER

SQUIGGLE MCZIGZAG
HOBBIES: DRAWING SQUIGGLY
MASTERPIECES, PLAYING
HIDE-AND-SEEK, AND DOING
SOMERSAULTS
FAVORITE FOOD: PICKLE
POPSICLES WITH KETCHUP DRIZZLE
FAVORITE COLOR: NEON GREEN

BINK WOBBLETOES
HOBBIES: BALANCING ON TIGHTROPES,
JUGGLING BOWLING BALLS, AND
TAP-DANCING
FAVORITE FOOD: PICKLE AND BANANA
SANDWICHES WITH CHOCOLATE SAUCE
FAVORITE COLOR: WOBBLY RED

SNICKER BUMBLEBOP
HOBBIES: SINGING SILLY SONGS,
TAP DANCING, AND PLAYING
THE KAZOO
FAVORITE FOOD: PICKLE ICE
CREAM WITH MUSTARD SWIRLS
FAVORITE COLOR: CHUCKLE PURPLE

FUZZY WHISKERPAWS
HOBBIES: POUNCING ON SHADOWS,
CHASING SQUIRRELS, AND NAPPING
IN SUNBEAMS
FAVORITE FOOD: MARSHMALLOW
PIZZA WITH ANCHOVIES
FAVORITE COLOR: MELLOW YELLOW

SPIKE SNORTSNOUT
HOBBIES: CLIMBING TREES, COLLECTING SHINY ROCKS, AND ROARING WITH LAUGHTER
FAVORITE FOOD: BARBECUE CHICKEN WINGS AND CRUNCHY POTATO CHIPS
FAVORITE COLOR: FIERY RED

WIGGLES MCTICKLETOES
HOBBIES: DANCING, TICKLING, AND PLAYING PRACTICAL JOKES
FAVORITE FOOD: JELLYBEANS AND GIGGLES
FAVORITE COLOR: RAINBOW

ZIPPY WHIRLWIND
HOBBIES: RACING WITH ROLLER SKATES, ZIPPING THROUGH TUNNELS, AND CATCHING FIREFLIES
FAVORITE FOOD: BANANA SPLITS WITH EXTRA SPRINKLES AND LIGHTNING BOLT-SHAPED COOKIES
FAVORITE COLOR: ELECTRIC YELLOW

GIZMO BUMBLEFOOT
HOBBIES: INVENTING GADGETS, BOUNCING ON POGO STICKS, AND SOLVING PUZZLES
FAVORITE FOOD: CHEESE DOODLES AND FIZZY SODA
FAVORITE COLOR: ELECTRIC BLUE

FUZZBALL FLUFFINGTON
HOBBIES: POPPING BUBBLE WRAP,
GIVING HUGS, AND COLLECTING SOCKS
FAVORITE FOOD: COTTON CANDY
CLOUDS AND BUBBLEGUM ICE CREAM
FAVORITE COLOR: FLUFFY PINK

ZIPPY WIGGLEPANTS
HOBBIES: SPEEDING ON ROLLER
COASTERS, EATING ICE CREAM
CONES, AND PLAYING AIR GUITAR
FAVORITE FOOD: MARSHMALLOW
HOTDOGS WITH KETCHUP
FAVORITE COLOR: TURBO ORANGE

ZIPPY WIGGLEPANTS
HOBBIES: SPEEDING ON ROLLER
COASTERS, EATING ICE CREAM CONES,
AND PLAYING AIR GUITAR
FAVORITE FOOD: MARSHMALLOW
HOTDOGS WITH KETCHUP
FAVORITE COLOR: TURBO ORANGE

SQUIGGLE NOODLEBRAIN
HOBBIES: DOODLING SILLY PICTURES,
MAKING UP FUNNY STORIES, AND
PLAYING TWISTER
FAVORITE FOOD: SILLY SPAGHETTI
FAVORITE COLOR: RAINBOW SWIRL

SNUGGLE FLUFFERNUTTER
HOBBIES: GIVING WARM HUGS,
BUILDING PILLOW FORTS, AND
STORYTELLING WITH FLUFFY CLOUDS
FAVORITE FOOD: MARSHMALLOW
SANDWICHES AND CUDDLY COCOA
FAVORITE COLOR: COZY BLUE

SNAPPY MCSNAP
HOBBIES: SNAPPING PICTURES,
TAKING SELFIES, AND MAKING VIDEOS
FAVORITE FOOD: SOUR GUMMY
WORMS AND POPCORN
FAVORITE COLOR: FLASHY YELLOW

WHISKERS MCSPRINKLETOES
HOBBIES: POPPING BUBBLE WRAP,
PLAYING WITH YARN BALLS, AND
CHASING LASER POINTERS
FAVORITE FOOD: TUNA FISH AND
SPRINKLE-COVERED CUPCAKES
FAVORITE COLOR: WHISKER PINK

FIZZY POPPER
HOBBIES: BUBBLE BLOWING,
PRANK PLAYING, AND HIDE-AND-
SEEK
FAVORITE FOOD: COTTON CANDY
AND POPCORN
FAVORITE COLOR: PINK

BIFF BEEFSTEAK
HOBBIES: LIFTING WEIGHTS,
FLEXING HIS MUSCLES, AND
SHOWING OFF HIS TATTOOS
FAVORITE FOOD: STEAK AND POTATOES
FAVORITE COLOR: ROYAL BLUE

FIZZY POPCORNOPOLIS
HOBBIES: MAKING BUBBLES,
PLAYING PRANKS, AND
SINGING SILLY SONGS
FAVORITE FOOD: COTTON
CANDY AND POPCORN
FAVORITE COLOR: BUBBLEGUM
PINK

GIZMO BISCUITBARREL
HOBBIES: COLLECTING SHINY OBJECTS,
BUILDING ROBOTS, AND PLAYING VIDEO
GAMES
FAVORITE FOOD: PEANUT BUTTER AND
JELLY SANDWICHES AND CHOCOLATE MILK
FAVORITE COLOR: ELECTRIC BLUE

BUBBLES BUBBLEWRAP
HOBBIES: POPPING BUBBLE WRAP,
BLOWING BUBBLES, AND SWIMMING
IN BUBBLE BATHS
FAVORITE FOOD: BUBBLE GUM AND
SODA
FAVORITE COLOR: SKY BLUE

THE CHOIR

FIZZLEWIGGLE MCJESTER
HOBBIES: TELLING JOKES, JUGGLING, AND DANCING
THE MONSTER MASH
FAVORITE FOOD: SILLY SPAGHETTI AND CHUCKLEBERRY PIE
FAVORITE COLOR: RAINBOW

DEAR FRIEND,

WE HOPE THAT THIS BOOK HAS BEEN A FUN AND ENJOYABLE WAY TO INTRODUCE YOUR LITTLE ONES TO THE WORLD OF COUNTING. COUNTING IS A FUNDAMENTAL SKILL THAT SETS THE FOUNDATION FOR SUCCESS IN MANY AREAS OF LIFE, AND WE BELIEVE THAT LEARNING SHOULD ALWAYS BE FUN AND ENGAGING.

BY USING MONSTERS AS THE SUBJECTS OF OUR COUNTING EXERCISES, WE AIM TO MAKE THE PROCESS BOTH ENJOYABLE AND MEMORABLE. WE HOPE THAT THE SILLY AND WACKY MONSTERS FOUND WITHIN THESE PAGES HAVE BROUGHT JOY AND LAUGHTER TO YOUR HOUSEHOLD.

REMEMBER, THE FUN DOESN'T HAVE TO STOP HERE. KEEP PRACTICING COUNTING WITH YOUR CHILDREN IN EVERYDAY SITUATIONS, AND ENCOURAGE THEM TO SEE THE WORLD THROUGH THE EYES OF CURIOSITY AND WONDER. WITH THIS ATTITUDE AND A SOLID FOUNDATION IN COUNTING, THE POSSIBILITIES FOR LEARNING AND GROWTH ARE ENDLESS.

THANK YOU FOR JOINING US ON THIS MONSTER COUNTING ADVENTURE!

BEST OF LUCK IN THE FUTURE FOR YOU AND YOUR LITTLE ONES!

MISTER JON
PANAJACHEL, GUATEMALA